Ladders

Animal Homes

World Book, Inc.
233 N. Michigan Ave.
Chicago, IL 60601
in association with Two-Can Publishing Ltd.

For information about other World Book publications, visit our
Web site http://www.worldbook.com or call 1-800-WORLDBK
(967-5325). For information about sales to schools and libraries,
call 1-800-975-3250 (United States); 1-800-837-5365 (Canada).

Written by: Sally Hewitt
Story by: Inga Phipps
Consultant: Sandi Bain
Main illustrations: Fiametta Dogi
Computer illustrations: James Evans
Editors: Julia Hillyard and Sarah Levete
Designer: Lisa Nutt
Managing editor: Deborah Kespert
Art director: Belinda Webster
Pre-press production manager: Adam Wilde
Picture researcher: Jenny West
U.S. editor: Sharon Nowakowski, World Book Publishing

2006 Printing
© Two-Can Publishing,1999

ISBN: 0-7166-7717-2
LC: 99-67036

Photographic credits: p4: Planet Earth Pictures; p6: Planet Earth Pictures; p7: Papilio Photographic;
p9: William Osborn/BBC Natural History Unit; p11: Oxford Scientific Films; p15: Planet Earth Pictures;
p16: Bruce Coleman Ltd; p18: Power Stock/Zefa; p20: Tony Stone Images; p21: David Fleetham/
Oxford Scientific Films; p22: Kit Houghton Photography; p23: Telegraph Colour Library.

Printed in China

6 7 8 9 10 09 08 07 06

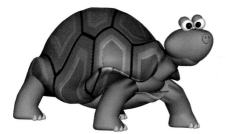

What's inside?

This book tells you about lots of different animal homes. You can find out where animals live—from high up in the treetops to deep underground—and how they build their amazing homes.

Holes and tunnels

Deep under the ground in the dark, moles, rabbits, and badgers live in long tunnels and snug holes. The big picture shows a mole hard at work, digging and scooping up earth for its new underground home.

A **mole** usually lives alone. It spends most of its time in the dark.

A rabbit's underground home is called a warren. When the coast is clear, the rabbit hops out and nibbles the grass.

A mole's underground tunnel is called a **burrow**. Inside, there are tasty worms for the mole to eat.

Strong **paws** shaped like shovels are perfect for moving earth.

It's a fact!

A badger is a busy housekeeper! On a warm day, the badger drags its bedding outside the burrow to air it in the sun.

A mound of earth piles up above the ground. This is called a **molehill**.

The mole kicks away earth with its **back legs**.

Baby moles sleep in a cozy room called a **chamber**.

 # Dens

Many animals make secret, hideaway homes in rocky caves and hollow tree trunks. These homes are called dens. In the big picture, a grizzly bear and its young are inside their warm cave.

A pile of **branches** keeps the cave hidden and warm.

In winter, this bear stays in its **cave**.

A young fox peeks out of its den in a tree trunk. In springtime, it's warm enough to come out and play!

The **rocky walls** of a cave keep out snow and wind.

Huge groups of bats live in caves. They sleep here all day long, hanging upside down. It's a tight squeeze!

A pile of leaves makes a soft **bed**.

A sleepy baby bear, called a **cub**, snuggles up to its mother.

Treehouses

It's safe high up in the trees!
Hungry enemies are far
below on the ground.
Squirrels build nests, while
a noisy woodpecker drills a
hole into a thick tree trunk.

It's a fact!

A tent bat camps out
in the trees! The bat
folds a leaf into a
tent shape, then
hides underneath.

A squirrel's **bushy tail**
helps it keep its balance
in the treetops.

These squirrels are
building their round **nest**
on a strong branch.

Soft **moss** keeps the inside of the nest warm and cozy.

Woven **twigs** and grass make the nest firm and strong.

A woodpecker taps at a tree with its long, sharp beak for hours. It makes a round hole, where it will live.

Logs and leaves

All kinds of tiny creatures make their homes on a mossy log. Here, they can sleep, eat, hunt, and lay their eggs. When the log becomes too crowded, the pile of leaves next door makes a safe place to shelter, too!

A shiny beetle crawls over the **moss**, looking for a crack where it can lay its eggs.

A beetle's eggs hatch into wriggling grubs that chew **wood** for their dinner.

Millipedes curl up under the **bark** where it is damp and dark.

A spider finds plenty to eat on the log. It spins a **web** to catch a juicy insect.

Weaver ants glue leaves together to make a nest. The queen ant, the most important ant of all, lays her eggs here.

This toad is cool and safe under a mound of fallen, decaying **leaves**.

A centipede is a fierce hunter. It chases spiders and beetles across the **log**.

In the woods

The woods are full of animal homes! Take a look high up in the trees, under piles of leaves, and near old, mossy logs.

12

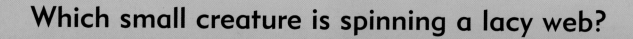

How does the noisy woodpecker drill a hole in the tree?

Words you know

Here are some words that you read earlier in this book. Say them out loud, then try to find the things in the picture.

molehill   den
bark   moss
twigs   web

How many bumpy molehills are popping out of the ground?

Birds' nests

In spring, many birds build comfortable nests, where they lay eggs. Soon the eggs hatch into hungry, noisy baby birds. The nest is home to the baby birds until they grow strong enough to look after themselves.

It's a fact!

Friendly weaver birds live in apartments! Together, they build a big grassy nest with a doorway for each bird.

A mother duck builds her **nest** near the edge of a lake.

She sits on her **eggs** to keep them warm.

Soft, fluffy **feathers** line the nest.

A stork builds a huge nest that it returns to every year. This stork's nest is perched on top of a tall chimney!

A baby duck is called a **duckling**.

Tall **reeds** of grass and cattails grow around the nest.

Hives and insect nests

A wasp nest is home to hundreds of striped wasps. Honeybees live together too, in homes called hives. Inside the hive, the bees make sweet honey that is delicious to eat.

Every week, a beekeeper visits the hive. He wears a mask and gloves to protect himself from the bees' sharp stings.

The queen wasp lays lots of tiny white **eggs**. •••••••

Each egg has its own special room, called a **cell**.

This **wasp nest** hangs down from the branch of a tree.

It's a fact!
A honeybee fills cells with honey. When a cell is full, the bee puts on a lid, just like putting a lid on a jar.

The eggs hatch into fat **grubs** that eat all day long!

Worker wasps chew wood to make **paper** to build the nest.

Lodges

A family of hard-working beavers help each other to build a home on the river. They use branches, sticks, and thick, brown mud. The big picture shows you what their home looks like above and below the water.

A beaver carries a **branch** in its mouth.

The family home is called a **lodge**. It makes a dam across the river.

A beaver gnaws at a tree with its strong front teeth. Before long, the whole tree falls over. *Crash!*

Sticky brown **mud** holds the branches firmly together.

Baby beavers stay snug and dry in a room above the water.

It's a fact!

A beaver nearly always knows which way a tree will fall. When the tree topples, the beaver dashes out of the way!

Beavers swim in and out of the lodge through a **tunnel**.

19

 # Shells

A few animals carry their homes around with them, wherever they go! A slimy snail has a shell on its back. In dry weather, the snail hides inside the shell, but when it's wet, it pokes out its head in search of tasty food.

A **snail** crawls along, searching for juicy leaves to munch.

A tortoise lives inside its tough shell. When the tortoise goes to sleep, it pulls its head underneath.

Feelers help the snail to find its way around.

A hard, round **shell** covers the snail's soft body. The shell is the snail's home.

A hungry **bird** is on .... the lookout for a tasty snail to eat.

When a snail is frightened, it pulls its **head** into its shell.

.... As a snail crawls along, it leaves a shiny, **gooey trail**.

A hermit crab doesn't have its own shell, so it borrows one! When the crab grows too big for the shell, it finds a new one.

Pets' homes

It can be great fun looking after a pet. There are lots of different kinds to choose from. Wild animals take care of themselves, but pets need people to give them food, water, and a clean, safe home.

Goldfish dart around and nibble on plants in a **tank** of fresh water.

A horse rests in a stable, but it also enjoys being outside in a field, where it can gallop and eat grass.

A hamster's home is like a playground with an **exercise wheel** and more!

A parakeet flies out of its **cage** to exercise.

An old rug in a basket makes a soft bed for a dog and an excellent place to hide bones!

In warm weather, guinea pigs scamper in an outdoor **pen**.

By the river

Many animals live by the river—in the trees, on the grassy banks, and even on the winding water itself.

24

Words you know

Here are some words that you read earlier in this book. Say them out loud, then try to find the things in the picture.

duckling   wasp nest
shell   branch
field   lodge

25

Lazy Beaver looks for a new home

Lazy Beaver woke up one morning to hear the sound of rain crashing down on the roof of the lodge. He looked outside and saw the river was about to burst!

Swoosh! Water swirled angrily around Lazy Beaver's feet as the river thundered along. The torrent raged, washing away the beavers' lodge completely.

Lazy Beaver grabbed a log and hung on for dear life as he was swept down the river. *Bump!* The log thumped into the riverbank and stuck fast among tangled tree roots.

The churning river was brown and thick with mud. Lazy Beaver crawled onto the bank and watched in dismay as the remains of the lodge floated downstream.

When the river had calmed, he noticed the other beavers from the lodge gathering on the opposite bank. One of the elder beavers was taking charge.

"We have to start building a new lodge!" cried the elder beaver. "The first job is to collect lots of branches and logs. Everyone start gnawing!"

The beavers scampered eagerly to gnaw and fell trees for logs and branches. "Come on, Lazy Beaver!" they called across the river, "help us build!"

"I'm fed up with building," grumbled Lazy Beaver to himself. "All we ever do is build and repair the lodge. And now that our old lodge is gone, we have to start all over again. I've had enough! I'm going to find myself a ready-made home." Lazy Beaver began to paddle slowly upstream.

"Quack!" said a duck as she swam past with her ducklings. The ducks climbed out of the water and into their feathery nest.

"That looks comfy," thought Lazy Beaver, "I could try living in a nest." He hid in the reeds and waited for the duck family to go for another swim. The mother duck soon splashed into the water and paddled away with her ducklings in a line behind her.

Lazy Beaver scrambled into the nest and flopped himself down. But it wasn't as comfortable as he had imagined. He was far too big for the ducks' nest. His paws hung over the edge, and his tail dragged in the water.

"Quack! Quack!"

Lazy Beaver looked up and saw the angry mother duck charging toward him. He quickly rolled over the side of the nest and into the water. *Splash!*

"Silly beaver! You've ruined my beautiful nest!" scolded the duck. "Go find your own home."

Lazy Beaver climbed out of the water and headed on his way. Nearby he saw a cave. "Now that looks like a nice home," he said, wandering inside. As his eyes got used to the dark, Lazy Beaver noticed hundreds of bats hanging from the ceiling.

"Wow!" thought Lazy Beaver, "those bats don't even have to build a home. They just find a cave and hang upside down all day. That's the life for me!"

Bending and stretching, Lazy Beaver tried to stand on his head. "Yikes! This isn't very comfortable," he panted, "especially when I start to wobble!"

Crash! Lazy Beaver lost his balance and landed on the cave floor in a furry heap.

"Silly beaver!" squeaked the bats. "You can't live here! Go find your own home."

Lazy Beaver rubbed his sore head and staggered outside into the bright sunshine.

BUMP! He immediately tumbled over a large shell on the ground. "I wonder what's in there," he said, peering underneath the shell.

From deep inside came a grumpy voice, "Do you mind?"

Then a tortoise's wrinkly head came slowly out of the shell. "I'm trying to sleep," said the tortoise.

Lazy Beaver looked at the shell curiously. "Does that shell make a comfortable home?" he asked.

"Of course it does, it fits me perfectly," replied the tortoise.

Lazy Beaver thought for a minute. "Please could I borrow it?" he smiled. "Our beaver lodge was washed away in the storm and I'm looking for a ready-made home."

The tortoise blinked a beady eye.

"Silly beaver!" he sniffed. "You can't borrow my shell. It grows on my back. Go find your own home!"

Weary, Lazy Beaver sat down at the bottom of a tree. "All I want is a comfy place where I can go to sleep. Surely it's not too much to ask?" he sighed miserably. As he wondered what to do, he gnawed at the tree.

Gnaw, gnaw, gnaw. Suddenly, the tree fell thundering to the ground. *Crash!* Lazy Beaver dodged out of the way. Then he climbed on to the felled tree trunk and closed his eyes. Soon he was fast asleep.

"Hey! Lazy Beaver!" shouted familiar voices. "Float that tree trunk down here."

Lazy Beaver woke up startled. The other beavers were calling to him from the new lodge down the river. "Come on, Lazy Beaver! Don't you want to live in the great new lodge?"

Lazy Beaver imagined curling up in a warm, dry lodge with his friends.

"Yes!" he shouted. "Of course I want to live in the new beaver lodge. I'm a beaver!" He rolled the tree trunk into the river and slipped in after it.

Lazy Beaver steered the tree trunk into place to complete the lodge, and all the other beavers clapped and cheered.

"Hooray! You're not such a lazy beaver after all. You finished building the lodge!"

But Not-so-lazy Beaver couldn't hear them. He was already snuggling down inside the splendid new beaver lodge for a nap.

Puzzles

Match the pairs!

Can you match the pairs of animals that live in the same kind of home? The questions below will help you.

rabbit

snail

bat

tortoise

bear

mole

Which two animals...

1 ...live in a burrow underground?

2 ...live inside a shell that they carry on their back?

3 ...might you find in a cozy cave?

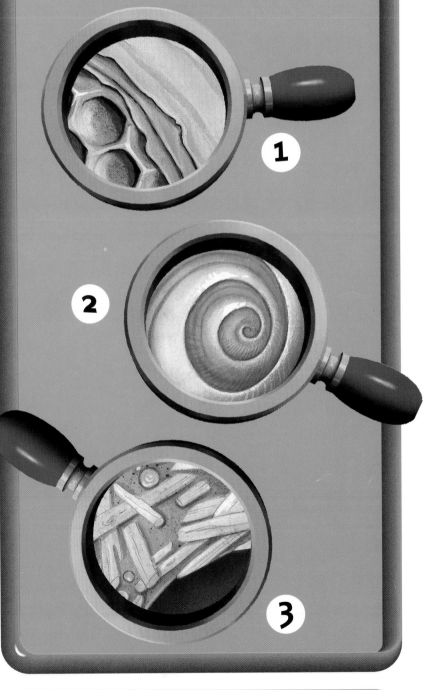

Close-up!

We've zoomed in on three different animal homes. Can you figure out which homes they are?

1

2

3

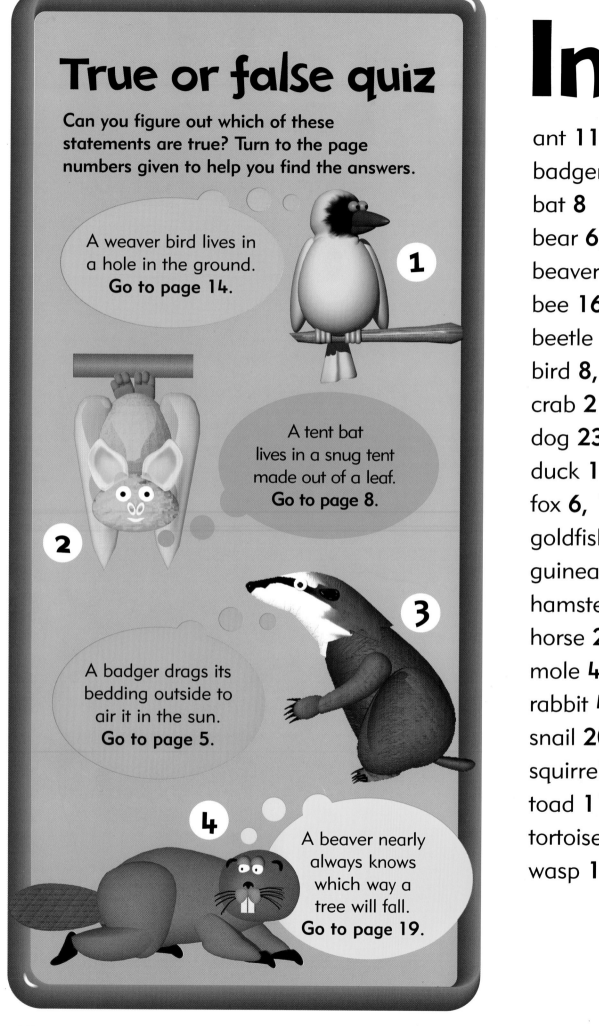

Index